1. Introduction

Many resource allocation mechanisms have the property that a number of individuals are vying for a potentially divisible, fixed resource. These individuals have costly actions available to them, which may influence the allocation of the resource depending on the allocation mechanism in place. In the literature, a common way to classify these mechanisms is whether the allocation depends randomly or deterministically on these actions.

A common example of a random allocation mechanism is a raffle, in which a single winning ticket is chosen from among a group of tickets purchased by many individuals. Moving away from this specific application, a broad variety of economic environments have been modeled as random allocation mechanisms. These include models of state lotteries, lobbying for government contracts, patent races, and competition for monopoly property rights. A common example of a deterministic mechanism is a first-price sealed bid auction in which a number of individuals submit bids, and the highest bidder buys the item for the amount he or she bid.

In the experimental economics literature there has been extensive investigation of deterministic environments, particularly auctions, and of one type of random allocation mechanism referred to as a single-prize contest. In the research reported here, we build upon the single-prize contest literature, and contrast behavior in this environment with another random allocation mechanism and a closely related deterministic mechanism.

We investigate a common single-prize contest in which individuals can make expenditures, and an individual's probability of winning the entire resource equals that individual's expenditure in proportion to the total expenditures of all individuals in the contest. The second random allocation mechanism we consider, the multiple-prize contest, is similar to the single-prize contest except the resource is divided into multiple prizes of fixed size, with individuals' expenditures influencing their probability of winning each of the prizes. The third allocation mechanism investigated is a deterministic mechanism in which each individual's share of the resource is that individual's expenditure in proportion to the total expenditures of all individuals in the contest. If all individuals are known to be risk-neutral, the normal form representations of our parameterizations of these three contests are identical.

Contrasting behavior in random and deterministic allocation mechanisms may be of interest to economists and policy makers for several reasons. In the literature on rent dissipation, the expenditures are typically viewed as being inefficient. On the other hand, as income generating mechanisms, these expenditures represent revenue. One complication when trying to select among these mechanisms, regardless of the objective, is that individuals' behaviors will likely depend on many factors including risk preferences. This is problematic for two reasons. First, individuals' risk preferences are not observable, nor are their beliefs about others' risk preferences. Second, as discussed later in this paper, economic theory shows that for some forms of contests, specific assumptions must be made about risk preferences in order to be able to rank the equilibrium level of expenditures. For example, only knowing that all individuals are risk-averse may not allow for such a ranking. Two of the contests we examine, the single-prize and multiple prize-contests, have this property. On the other hand, a nice property of the deterministic contest we examine, the proportionate prize contest, is that regardless of risk preferences of individuals, it imposes incentives that are identical to those that would be present if everyone were risk-neutral in the single prize contest.

Section 2 of this paper describes the experimental decision environment we designed to investigate these mechanisms. Section 3 presents formal models of the specific mechanisms used in the experiments, and investigates equilibrium behavior under risk-neutrality and in its absence. Section 4 presents the results of the experiments. Section 5 briefly discusses how these results relate to some common explanations of diverse behavior in economic environments. Section 6 concludes.

2. The Decision Setting

The study is based on two experimental sessions, with a total of 44 participants, which were conducted one week apart during Spring semester 2001. Participants were recruited from introductory economics classes at Indiana University-Bloomington.[1] Upon entering the classroom used to conduct the experiment, participants were anonymously assigned to four-person groups. Participants knew that three of the roughly twenty other people in the room were in their group, but had no idea which three.

Participants were presented with instructions that summarized the experimental procedures, and gave a detailed description of each of the three decision situations. These instructions are included in the appendix. Instructions were read by each participant in private and reviewed publicly by the experimenter, using an overhead projector. Participants were also presented with a demonstration of the procedures for determining the outcome of each situation. Fictitious, randomly chosen, decisions were used for the demonstration. Upon reviewing the instructions for each situation, participants were given time to make their decision for that situation by filling out a form, in pencil. After all three situations had been presented, participants were given an opportunity to review all of their own decisions and make any changes they desired. Participants were not allowed to observe the decisions made by any other subject until after all the decisions had been collected.

It was public information that after decisions were collected, the experimenters would randomly choose one decision situation for each four-person group to determine experimental earnings. The choice of decision situation for each four-person group was made in public by blindly picking one of three color-coded chips from a cup. The chosen situation was then played out in public view and participants' earnings for the experiment were calculated. Each participant also received a $5 participation fee. Each session lasted less than an hour. Participants' earnings ranged from $5 to $94, with a median of $25.

The three decision situations were presented to the participants as: the first decision situation, the second decision situation, and the third decision situation. For purposes of discussion, we refer to the three situations as: the single-prize contest, the multiple-prize contest, and the proportionate-prize contest, which represents the order of presentation to the participants. As was the case in the instructions, the

---

[1] Students in introductory economics have majors in numerous disciplines including business, political science, journalism, and economics. Less that 5% are economics majors.

descriptions presented below are from the perspective of a particular contest being randomly chosen as the one on which experimental earnings would be based.

The single-prize contest was presented as a situation in which one of the four participants in a group received a prize of $72. Each participant was given an endowment of $20 and could affect the likelihood of receiving the prize by purchasing tickets for 25¢ each. The recipient of the $72 prize was chosen by randomly drawing a ticket from among all of the tickets purchased by a group. It was public information that the prize would be awarded randomly if none of the participants in a group purchased a ticket, with each participant having an equal chance of receiving the prize. The winning ticket was chosen using the visual tool of a computerized "spinning wheel," where tickets purchased by each participant in a group were numbered, and those numbers were displayed in random order on a wheel. The wheel was programmed to spin for a random duration of time, at which point the winning number was displayed. A participant's earnings equaled the $20 endowment, minus ticket purchase costs, plus $72 if they were the winner.

The procedures for conducting the multiple-prize contest were identical to those for the single-prize contest, except that there were three $24 prizes, with each participant having the potential to win one, two, or all three prizes. Again, participants were given an endowment of $20 and tickets cost 25¢ each. It was public information that each prize would be awarded randomly if none of the participants in a group purchased a ticket, with each participant having an equal chance of winning each prize. The outcome of this contest was determined by the spinning wheel described above displaying three winning tickets when it stopped. A participant's earnings equaled the $20 endowment, minus ticket purchase costs, plus $24 for each of the three prizes they won.

The proportionate-prize contest was presented as an opportunity for the participants within a group to receive a share of a $72 prize. Identical to the other two situations, participants were endowed with $20 and participants could purchase tickets at a cost of 25¢. The share of the prize received by each participant equaled the proportion of his/her tickets relative to those of the entire group. It was public information that if no participant in a group purchased tickets, each received ¼ of the prize. Earnings by a participant equaled their $20 endowment, minus ticket purchases, plus their share of the $72 prize.

3. Formal Models of the Mechanisms

Single-Prize Contest

The single-prize contest model used in this paper was originally developed by Tullock (1980). Using the parameters from our experiment, participant i's expected payoff is

$$E\pi_i(X) = \left(\frac{x_i}{\sum_{j=1}^{4} x_j}\right) u_i\left(\$20 + \$72 - \frac{\$x_i}{4}\right) + \left(\frac{\sum_{j \neq i} x_j}{\sum_{j=1}^{4} x_j}\right) u_i\left(\$20 - \frac{\$x_i}{4}\right) \quad \forall i = 1, 2, 3, 4 \quad (1)$$

In this expression, $X=(x_1, x_2, x_3, x_4)$ denotes the vector of the number of tickets purchased by each participant, and $u_i()$ represents i's utility for the change in income due to the experimental earnings. If all participants are risk-neutral, then the Nash equilibrium purchase is $x_i=54$.

Multiple-Prize Contest

In the multiple-prize contest, there are three prizes of $24 each. The prizes are awarded by drawing three tickets without replacement from amongst all the tickets purchased within a group. Thus, the expected payoff of participant i is given by:

$$E\pi_i(X) = \sum_{k=0}^{3} \left(\frac{C(x_i, k) \, C\left(\sum_{j \neq i} x_j, 3-k\right)}{C\left(\sum_{j=1}^{4} x_j, 3\right)}\right) u_i\left(\$20 + \$24k - \frac{\$x_i}{4}\right) \quad \forall i = 1, 2, 3, 4, \quad (2)$$

where $C(a,b)$ is the number of unordered groups of size b that can be selected from a set of size a, i.e., it is a binomial coefficient. If all participants are risk-neutral, this simplifies to equation (1). Thus, the Nash equilibrium number of tickets purchased in the multiple-prize contest is also 54 if all players are risk-neutral.

Intuitively, the probability of a marginal ticket being selected to win a specific one of the three prizes is one divided by the total number of tickets purchased. Thus, the chance of winning any of the three prizes with a marginal ticket is three over the total number of tickets. Therefore, the number of dollars the participant would expect to win by purchasing that ticket is $24 times 3 over the total number of tickets, which is $72 over the total number of tickets, which is exactly the expected number of dollars a participant would win by purchasing an extra ticket in the single-prize contest. Since the marginal expected gain of a ticket is the same in this contest as in the single-prize contest, and the marginal cost of a ticket is the same, the best response functions, and thus the Nash equilibrium, must also be the same for a risk-neutral participant.

Proportionate-Prize Contest

In the proportionate-prize contest, the $72 prize is divided among the participants in amounts proportionate to the number of tickets they purchase. The payoff to participant *i* is just his utility from the endowment plus his share of the prize minus the cost of his tickets.

$$\pi_i(X) = u_i\left(\$20 + \$72\frac{x_i}{\sum_{j=1}^{4} x_j} - \frac{\$x_i}{4}\right) \quad \forall i = 1,2,3,4 \qquad (3)$$

Notice that there is no need to take an expectation, because this mechanism is deterministic. Also note that the argument of the utility function in (3) is exactly the expected payoff for a risk-neutral participant in the single-prize contest as specified in equation (1). Thus, the Nash equilibrium number of tickets in the proportionate-prize contest is $x_i = 54$.

Equilibrium Implications

If all participants were risk-neutral expected utility maximizers, and this was common knowledge, then the unique Nash equilibrium to all three of the contests described above would be for each participant to purchase 54 tickets. We see this common equilibrium prediction as being one of the several dimensions in which these three contest mechanisms are comparable. The other important similarities are that the participants' available strategies are identical in all three contests, and that the total amount of prize money (rent) being awarded is the same in all three.

Upon first glance, one might conjecture that risk-averse participants would be most hesitant to purchase tickets in the single-prize contest because it involves a lottery over two relatively extreme outcomes. On the other hand, the only uncertainty in the proportionate-prize contest is due to the strategic uncertainty concerning ticket purchases by others in one's group. This uncertainty is present in all three contests, so the purchase of a ticket in the proportionate-prize contest would seem to be less risky than in the other two contests. Consequently, one might expect relatively risk-averse participants to purchase the most tickets in that contest. The multiple-prize contest lies somewhere between these other two. It has uncertainty due to the random selection of winning tickets; however, the division of the prize money into three parts makes it a somewhat less extreme lottery than the single-prize contest.[2]

If participants have diverse risk preferences and diverse beliefs about others' risk preferences, neither of which can be observed or controlled, then nearly any behavior could be consistent with a

---

[2] This argument about the relative uncertainty surrounding the equilibrium payoffs of the three contests can be substantiated, somewhat, by comparing the variances in expected payoffs across the three contests in equilibrium. For instance, in the risk-neutral Nash equilibria, all three contests yield expected payoffs of $24.50. In the proportionate prize contest, this payoff is obtained with certainty in equilibrium, so the equilibrium payoff has no variance. In the multiple prize contest, the variance of the expected risk-neutral Nash equilibrium payoff is roughly 321. This is much lower than in the single prize contest, where the variance is 972. Of course, this does not necessarily indicate how risk averse agents would rank the desirability of the three contests, much less how their risk preferences would change their equilibrium play. Absent specific information about how the equilibria change when certain types of risk preferences are present, this information is simply provided as one potential indication that players may view the single prize contest to be the most uncertain and the proportionate prize contest to be the least uncertain. However, as will be discussed briefly, the presence of risk aversion alone is not sufficient to generate any consistent prediction about the relative equilibrium actions in these three contests.

Bayes-Nash equilibrium to an appropriately specified model (Ledyard 1986, Börgers 1993). In fact, Konrad and Schlesinger (1997) have shown that the direction of the change in equilibrium ticket purchases in the single-prize contest caused by an increase in risk aversion is indeterminate. Since the proportionate-prize contest imposes a reward structure in which utility maximizing participants should act as if they were risk-neutral participants in a single-prize contest, the Konrad and Schlesinger (1997) result implies that even if it was known that participants were homogeneous and risk-averse, there is no theoretical foundation for predicting the relative magnitudes of ticket purchases in the single-prize contest versus the proportionate-prize contest.[3] Given the indeterminacy in the theoretical literature, and the growing prevalence of winner-take-all mechanisms documented by Frank and Cook (1995), we seek to examine the impact of these various reward schemes on rent-seeking behavior experimentally.[4]

4. Results

This section presents summary observations for all three contests. It then focuses on how individual decisions vary across pairs of contests.

---

[3] Despite the indeterminacy of the impact of risk aversion in general, we have verified that if all subjects in our contests are risk averse, and have identical utility functions displaying either constant relative risk aversion, or constant absolute risk aversion, then equilibrium ticket purchases are greatest in the proportional prize contest, followed by the multiple prize contest, then the single prize contest. This ordering also holds if all players have utility functions of the form: $u(x)=x/(x+b)$, where an increase in the positive parameter b increases the Arrow-Pratt measure of relative risk aversion and decreases the Arrow-Pratt measure of absolute risk aversion. We know of no utility function which would alter the ordering of equilibrium ticket purchases across our three contests. However, we have verified that in a two player version of the contests considered here, the equilibrium ticket purchases in the single prize contest would be larger than in the proportionate prize contest if both players had utility of the form $u(x)=ArcTan(x/80)$. This is a concrete example of Konrad and Schlesinger's (1997) result.

[4] See Frank and Cook (1995) for an intriguing discussion of the evolution of markets towards "winner take all" contests, especially the markets for highly skilled labor.

Summary Observations

Figure 1 shows the cumulative frequency of ticket purchases in each of the three contests. One observation that is immediately apparent is that there is considerable variation in the number of tickets purchased in each of the three contests. There are mass points for each contest, such as 20 tickets for the single-prize contest. However, ticket purchases vary from 0 to 80 in all three contests. While purchases in the single-prize contest are the lowest of the three on average, and purchases in the proportional-prize contest are the highest, differences in the distribution of ticket purchases across the three contests are not very large.

No previously run experiments are sufficiently closely related to either the multiple- or proportionate-prize contests to allow for comparisons of results. There are, however, experiments from the rent seeking literature, including those by Millner and Pratt (1989, 1991), Shogren and Baik (1991), Davis and Reilly (1998), Potters et al (1998) and Önçüler and Croson (1998), which investigate single-prize contests under various experimental settings. All of these settings, however, involve repeated play of a single-prize contest and are therefore not truly one-shot as in this study.[5] The single-prize contest experiment most closely related to the one presented here is by Shupp (2000). This experiment involves four-person groups competing for a single prize of the same dollar amount used here.[6] It also involves repeated play (with a random matching protocol), but a meaningful comparison can be made by focusing on the first round. Average first round purchases were 37% of the risk-neutral Nash prediction in Shupp's experiments, considerably lower than the 70% of the risk-neutral Nash prediction averaged in this study. It should be noted that average purchases in Shupp (2000) increased over the course of the experiment to

---

[5] Several of these experiments use random matching protocols to simulate a one-shot setting. See for example Potters et al (1998).

[6] This study also involved multiple-prize contests, but they differ from the multiple-prize contest in this study by allowing only one prize per participant.

91% of the risk-neutral Nash prediction, and that ticket purchases varied widely across individuals throughout all rounds.

It is possible that subjects in our experiments would have changed their behavior if they had played these contests repeatedly. Such an observation would raise at least as many questions as it would answer. Would it be seen as evidence that subjects learn how the allocation mechanisms work? If so, repetition would seem to provide a better test for static equilibrium predictions. However, subjects may also learn about how others play the game. This may change their assumptions about how others will play in the future. It may also make them reconsider the way they were thinking about the contests. Neither of these effects are desirable when testing a static equilibrium concept. Schmidt, Shupp, Walker, and Ostrom (2003) used experimental evidence from one-shot and repeated play of simple bi-matrix coordination games to investigate this tradeoff. They found that the subset of data from their repeated game environments that most closely resembled their one-shot data came from the first round of experiments in which subjects were randomly rematched each period. This data was not significantly different than the one-shot data, whereas data from last round of the randomly rematch experiments, and the first and last rounds of experiments in which subjects remained matched with the same subject for the entire experiment, were significantly different than the one-shot data. Given the simplicity of a bi-matrix game, this result raises a question about the practice of using a sequence of games against randomly selected opponents as a proxy for one-shot games. Since most people are very familiar with the notion of a raffle, we did not feel the contests studied here were so complicated that the risk of introducing repeated-game effects into our results was worth the potential benefit of having people become more familiar with our contests. The goal was to design an environment consistent with the conditions of a static game, not one that would lead to Nash play.

To examine how individuals view the three contests relative to one another, we classify participants based on how their ticket purchases in each contest rank relative to their purchases in the other contests. For instance, one classification is for participants who purchase at least as many tickets in the single-prize contest as in the multiple-prize contest, and at least as many in the multiple-prize contest as in the proportionate-prize contest. This classification would be referred to with the notation: s≥m≥p. The relative frequencies of participants in each of the seven classifications are given in Table 1.[7]

We can see that the most frequently observed category is one in which the participants purchased the most tickets in the proportionate-prize contest and the least in the single-prize contest (p≥m≥s), which accounted for 27% of the participants. The second most frequently observed category is one in which the participants purchased the same number of tickets in all three contests (p=m=s), which accounted for 21% of the observations. Even within this category, there was considerable variation. For example, two participants purchased zero tickets in all three contests, two more purchased 20 in all three contests, and two more purchased 80 in all three contests. The category (p≥s≥m) was the third most frequently observed at 19%. So in total, 67% of all participants bought at least as many tickets in the proportionate-prize contest as in either of the other two contests. But what is most clear from this analysis is that there is a great deal of heterogeneity in subjects' actions in and across these contests.

Paired Contest Comparisons

Single- versus Multiple-Prize Contests

There is strong evidence that participants treat the single- and multiple-prize contests as being very similar. The mean number of tickets purchased in the single-prize contest is 37.6, and the standard

---

[7] We use weak inequalities in this analysis to keep the number of categories relatively small. In analysis not presented here, we have also classified the data using strict inequalities and equalities. For instance, the category s≥m≥p in the analysis in the paper could be split into categories s>m>p, s>m=p, s=m>p, and s=m=p. It should be noted, that if we have an observation in which s>m=p, that observation would fit into two of our categories, s>=m>=p and s>=p>=m. For the relative frequencies reported in Figure 2, such an observation would be counted as half an observation in each of these two categories. Finally, given the large number of subjects (9 out of 44) who purchased the same number of tickets in all three contests, we have set up a separate classification s=m=p. Any observation consistent with this classification is only counted in this classification, even though such an observation would also be consistent with every other classification.

deviation is 23.0. In the multiple-prize contest, the mean is 39.5 and the standard deviation is 22.5. A two-sided t-test assuming unequal variances fails to reject that these observations came from distributions with equal means (p-value=0.70). This simply demonstrates that ticket purchases are on average similar, but it does not imply that this carries over to the individual level.

The first panel of Figure 2 displays each individual's ticket purchases in the single- and multiple-prize contests. A data point represents an individual's ticket purchases in the single-prize (x-axis) and the multiple-prize (y-axis) contests. The number at each point equals the number of individuals with the corresponding purchases. Consistent with the apparent correlation in the first panel, the coefficient of correlation between single-prize and multiple-prize ticket purchases is 0.80. Furthermore, 20 of the 44 participants purchased identical numbers of tickets in the two contests. Nine participants bought more tickets in the single-prize contest than in the multiple-prize contest, and fifteen subjects bought more in the multiple-prize contest.

Single- versus Proportionate-Prize Contests

In the proportionate-prize contest, the mean ticket purchase is 44.4 and the standard deviation is 25.7. When comparing this data with the single-prize observations, a two-sided t-test assuming unequal variances fails to reject that these observations came from distributions with equal means (p-value=0.19). The significance level of this test indicates that there may be somewhat less similarity in how participants behave in these two contests than in the previous comparison.

The second panel of Figure 2 displays individuals' decisions in these two contests. Notice that there appears to be relatively more dispersion away from the forty-five degree line than there was in the first panel. This lack of correlation is further evidenced by the coefficient of correlation between single-prize contest ticket purchases and proportionate-prize contest ticket purchases, which is 0.31. Furthermore, only eleven participants purchased identical numbers of tickets in these two contests. The majority of participants, 23 of 44, purchased more tickets in the proportionate-prize contest than in the single-prize contest. Ten participants purchased fewer tickets in the proportionate-prize contest.

Multiple- versus Proportionate-Prize Contests

Using the means and standard deviations reported above, a two-tailed t-test fails to reject that the mean ticket purchases are equal in these contests (p-value=0.34). The third panel of Figure 2 indicates that there is very little correlation in ticket purchases in these two contests, and that is confirmed by the coefficient of correlation, which is 0.29. Again, roughly one-fourth of the subjects (12 of 44) purchased the same number of tickets in these two contests. Close to half of the participants (20 of 44) bought more tickets in the proportionate-prize contest, while twelve participants bought more tickets in the multiple-prize contest.

5. Exploring Dispersion

A quick glance at Figure 1 indicates that one would have to reject any equilibrium model that predicted identical actions by all subjects in any of the contests based on our data. The risk-neutral Nash prediction is for everyone to buy 54 tickets in each of the three contests, but none of the subjects actually purchased 54 tickets in any of the contests, and ticket purchases varied from 0 to 80 in all three. One explanation for this is that this situation represents disequilibrium. That hypothesis is impossible to reject, given that we cannot observe subjects best response functions. We would like to see if this behavior can be explained by certain commonly assumed types of individual heterogeneity, or alternate equilibrium concepts.

Often when subjects' behaviors differ from the risk-neutral Nash prediction, it is suggested that other risk preferences might be more appropriate.[8] Allowing for heterogeneous risk preferences could explain the dispersion in the single prize and the multiple prize contests, but not in the proportionate prize contest. In the proportionate prize contest, as long as all players are expected utility maximizers, the unique Bayes-Nash equilibrium is for all players to buy 54 tickets regardless of assumptions about distributions of risk preferences.

---

[8] For example, Cox, Smith, Walker (1985) consider risk aversion in experimental first-price sealed bid auctions. Also, Holt and Laury (2002) experimentally test for risk preferences in simple gambles.

Another possible explanation for this dispersion comes from relatively new equilibrium concepts which have recently gained popularity, in which players systematically tremble from their best responses (Beja 1992, McKelvey and Palfrey 1995). For instance, certain specifications of McKelvey and Palfrey's quantal response equilibrium produce equilibrium mixed strategy profiles in which play is qualitatively consistent with the patterns of play shown in Figure 1. However, if the dispersion of ticket purchases shown in Figure 1 is explained simply by players' actions being based on different draws from common equilibrium mixed strategies for each contest, then players' actions should be uncorrelated across contests. The first panel of Figure 2, for instance, suggests there is a positive correlation between ticket purchases in the single and multiple prize contests. Therefore, noisy best responses alone will not explain the pattern of ticket purchases observed across all three contests.

Taken together, heterogeneous risk preferences and noisy best responses can explain the pattern of play we see in and across the three contests, with one major exception. Figure 1 clearly depicts that the most frequently observed numbers of tickets purchased are multiples of 20, accounting for over 73% of all purchases.[9] Since tickets cost $0.25 each, the prevalence of ticket purchases that are multiples of twenty may suggest that many subjects selected to work in $5 increments. This might indicate that subjects are attempting to simplify a complicated decision environment by viewing the strategy space with a coarser grid. To solve for a quantal response equilibrium with homogeneous players, one must solve a system of equations in which the number of equations and unknowns is equal to the number of potential actions players have. So, to find a quantal response equilibrium if players were allowed to buy any integer number of tickets between 0 and 80, one would need to solve a system of 81 non-linear equations in 81 unknowns. Adding player heterogeneity makes matters worse. For each player type added to the model, an additional 81 equations and 81 unknowns enter the system. This clearly becomes infeasible rather quickly, if not immediately. Since the players themselves appear as if they might be

---

[9] Forty was the most common number of tickets purchased, accounting for 21% of all observations, followed by 20 (19%), 80 (14%), 60 (13%), 0 (6%), and 28 (4%).

simplifying the game by reducing the strategy space through the use of a coarser grid, we decided to adopt the same approach for this analysis. The analysis that follows will consider only ticket purchases that are even multiples of twenty, so that each player type adds just five equations and five unknowns.

In a Nash equilibrium, players must only put positive probability on playing strategies that are best responses to other players' equilibrium strategies. In a quantal response equilibrium, players play mixed strategies in which the probability of playing a strategy is increasing in the expected utility of that strategy given that others are playing the equilibrium. Exactly how the mixed strategies must relate to the expected payoffs is part of the equilibrium specification. One particular specification of a quantal response equilibrium maps expected payoffs to mixed strategies using a logistic function, hence this specification is referred to as the logit equilibrium. Let σ_{-i} be the vector of mixed strategies employed by all players other than player i, and $U_i^e(x_i, \sigma_{-i} | \theta_i)$ be player i's expected utility from buying x_i tickets when all other players are playing according to σ_{-i} and player i is of type θ_i. Player i's mixed strategy, σ_i, is a logit best response if:

$$\sigma_i(x_i | \theta_i) = \frac{\exp(U_i^e(x_i, \sigma_{-i} | \theta_i)/\mu)}{\sum_{j=0}^{4} \exp(U_i^e(20j, \sigma_{-i} | \theta_i)/\mu)}$$

for x_i=0, 20, 40, 60, and 80, where $\mu>0$ is a parameter that essentially describes the level of noise in the decision making process. If μ is very large, then each x_i will be played with a probability close to 1/5. If μ is close to zero, then the strategy with the highest expected payoff will be played with a probability close to one, approximating the strict best response required by the Nash equilibrium. Therefore, lower values of μ can be thought of as representing lower levels of noise. The mixed strategy profile (σ_i, σ_{-i}) is a logit equilibrium if every type of every player is playing a logit best response as defined above. As discussed above, if all players were identical, players' actions would not be correlated across the three contests.

To get correlation of actions across contests in equilibrium, we will solve for a logit equilibrium in a model where players' utility functions for money, the u_i's from equations (1) through (3), are drawn from a distribution of constant relative risk aversion utility functions of the form: $u_i(y) = y^{\theta_i}$. The logit equilibrium can then be calculated for a given set of parameters: the potential θ's; a probability distribution over that set; and a value of μ for each contest. Thus, if there are k types of players, the parameter vector is (θ^1, θ^2, ..., θ^k, p^1, ..., p^k, μ_1, μ_2, μ_3). One reason to estimate a separate value of μ for each contest is that the complexity of the decision environment, as seen by the subjects, may vary across contests. This perception of complexity may influence subjects' willingness to expend effort to determine an optimal strategy, which could affect the level of noise in the environment.

For each specific parameterization, a logit equilibrium strategy profile places a positive probability on each possible vector of ticket purchases across the three contests, so one can calculate the likelihood of observing the set of observations that came from our experiments by multiplying together the probabilities with which each subject's actions would be observed in equilibrium. As noted above, the computational complexity of the equilibrium calculation is increasing in the number of player types. We report here the results of a likelihood-based estimation procedure in which the number of player types was taken to be four.[10] We do not see the choice of parameterization as being terribly important because we simply use this procedure to find a parameterization that yields an equilibrium that is qualitatively similar to our experimental data. The possibility that another parameterization may better fit our data is thus of no concern. It is entirely possible that allowing for heterogeneity across individuals in the noise parameters and functional forms of utility functions would provide a better fit. Given the computational complexity

---

[10] With four player types, we have 11 parameters. For each specification of this parameter vector, solving for an equilibrium involves finding a root of a system of 25 equations in 25 unknowns. We also tried allowing for five and six types of players, but the extra types did not allow the estimation procedure to reach a substantially higher estimated likelihood. Given the complexity of the likelihood function, we cannot be sure that our hill climbing maximization routine is actually reaching a global maximum, it may simply have found a local maximum, so we do not refer to this as a maximum likelihood estimate. However, visual inspection of grids of the likelihood function suggest that it is relatively well behaved. Furthermore, we tried starting this hill climbing procedure with many diverse starting points, which all led to this same maximum.

of our likelihood function, we do not believe ours is a good application to explore such matters. We simply wish to demonstrate that even though neither heterogeneous risk preferences nor noisy best responses are sufficient to generate equilibria consistent with our experimental data, the two combined are.

This estimation procedure yielded four coefficients of relative risk aversion: (0.03, 0.27, 0.75, and 2.53). The estimated distribution of subjects across these four types is given by: (0.13, 0.24, 0.38, and 0.25). For instance, this says that 13% of the people are estimated as having a utility function of the form: $u_i(y) = y^{0.03}$. The estimated noise parameter for contest three is considerably larger than the noise parameters for the other two contests. The estimation procedure yields $\mu_1=0.002$, $\mu_2=0.008$, and $\mu_3=0.032$. Diverse risk preferences alone are enough to generate diverse actions in the first two contests, but the noisy best responses are necessary for dispersion in equilibrium actions in the third contest.

Since we view this as primarily being an illustrative example, we will focus on a comparison between the observed relative frequencies of outcomes in our experimental setting to the frequencies from an equilibrium derived from the likelihood based procedure discussed above. Since we wish to explain both dispersion within each contest and correlation across contests, we present a comparison that addresses both of those issues. Figure 3 is a bubble chart which focuses on this comparison for the single prize contest and the multiple prize contest. At the coordinate (0,0), there is a number 3 printed on a very small shaded circle surrounded by the outline of a larger circle. The number 3 indicates that three subjects bought zero tickets in both the single prize contest and the multiple prize contest (when ticket purchases are rounded to the nearest multiple of 20). The size of the larger circle that outlines the number and the shaded circle is indicative of this frequency, where the diameter of the circle is proportional to the observed frequency. The small shaded circle at the point (0,0) is indicative of the frequency that individuals would purchase zero tickets in each of these contests in the equilibrium. If the equilibrium had predicted that 3 out of 44 subjects would play (0,0), the shaded circle would have been the same size as the outline circle. So this figure shows us that more subjects bought no tickets in both of these contests than would be predicted in equilibrium. By contrast, no number or outlined circle appears at (20,0), indicating that no subjects actually bought twenty tickets in the single prize contest and zero tickets in the multiple prize contest. However, there is a small shaded circle, indicating that (20,0) occurs with positive probability in equilibrium. At the point (20,40), the shaded circle and the outline circle are approximately the same size, indicating that the observed frequency of this outcome is nearly the same as the equilibrium frequency.

Since equilibria are mixed strategy profiles, and observed frequencies would have to be considered realizations of random variables, we cannot simply look for discrepancies between the two types of circles to determine whether the equilibrium is failing to describe the data because we only have 44 observations of the experimental data. We can, however, see whether the qualitative properties of the equilibrium predictions are consistent with the observed frequencies. Ticket purchases in both contests range from 0 to 80 in both the equilibrium and the experimental data. Also, ticket purchases are positively correlated across the single prize and multiple prize contests in both the equilibrium and the experiment.

Figures 4 and 5 present similar comparisons for single prize versus proportionate prize, and multiple prize versus proportionate prize. Less correlation was seen across contests in both of these cases, and the equilibrium captures that. The equilibrium frequencies show less correlation in these figures than was seen in Figure 3 because the dispersion in proportionate prize actions is driven more by the noise parameter, μ_3, than in the other contests. The distribution of risk preferences played a more important role in the other two contests. An individual's risk preference stays the same across the three contests, thus their actions will tend to be correlated across contests in which risk preferences are important. Noise is, by definition, an independently and identically distributed random variable, and is not correlated with risk preferences.

6. Summary and Conclusions

We investigate the resource allocation properties of three institutions: a probabilistic single-prize contest, a probabilistic multiple-prize contest, and a deterministic proportionate-prize contest. In the first two institutions, an individual's probability of receiving a prize is proportionate to their contest expenditures relative to the expenditures of other contest participants. In the third institution, an individual's share of the prize is proportionate to their contest expenditures relative to expenditures of other contest participants. If all participants are risk-neutral, the predicted level of expenditures is equivalent for all three institutions. In summary, for the experiments reported here, expenditures within each type of contest vary considerably across participants, on average falling below the predicted level of expenditures for the risk-neutral benchmark. In addition, there is evidence that participants tend to make larger expenditures in the deterministic proportionate-prize contest compared to the probabilistic contests.

One interpretation of expenditures in contests of the types investigated in this study is that of rent dissipation. Participants may be viewed as competing for an exogenously available resource. Expenditures used to capture the resource are wasteful from the perspective of the social optimum. An alternative interpretation is from the perspective of the revenue generating properties of the three institutions. To the extent that the institutional form of resource allocation mechanisms are endogenously determined, studies of the type presented here provide a testing ground for investigating the allocative and income generating properties of alternative institutions. In this regard, the results reported here suggest that institutions based on deterministic sharing rules for fixed prizes may lead to greater levels of rent dissipation, or alternatively, greater revenues to those conducting the contests, than random sharing rules for a given subject pool. However, it is easy to imagine situations in which this result would not hold if individuals were allowed to self-select across the set of potential mechanisms.

References

Beja, A. (1992). "Imperfect equilibrium," *Games and Economic Behavior*, 4:18-36.

Börgers, T. (1993). "Pure Strategy Dominance," *Econometrica*, 61:421-430.

Cox, J. C., Smith, V. L. and Walker, J. M. (1985). "Experimental Development of Sealed-Bid Auction Theory; Calibrating Controls for Risk Aversion," *The American Economic Review,* 75(2):160-165.

Davis, D. and Reilly, R. (1998) "Do Too Many Cooks Always Spoil the Stew? An Experimental Analysis of Rent-Seeking and the Role of a Strategic Buyer", *Public Choice,* 95(1-2): 89-115.

Frank, R. H. and Cook, P. (1995). *The Winner-Take-All Society.* Penguin Books, New York, New York.

Holt, C. A. and Laury, S. K. (2002), "Risk Aversion and Incentive Effects," mimeo.

Konrad, K. A. and Schlesinger, H. (1997). "Risk Aversion in Rent-Seeking and Rent-Augmenting Games", *Economic Journal*, 107:1671-1683.

Ledyard, J.O. (1986). "The Scope of the Hypothesis of Bayesian Equilibrium," *Journal of Economic Theory*, 39:59-82.

McKelvey, R. and Palfrey, T. (1995). "Quantal Response Equilibria for Normal Form Games," *Games and Economic Behavior*, 10:6-38.

Millner, E. and Pratt, M. (1989). "An Experimental Investigation of Efficient Rent-seeking", *Public Choice*, 62(2), 139-151.

Millner, E. and Pratt, M. (1991). "Risk Aversion and Rent-seeking: An Extension and Some Experimental Evidence", *Public Choice*, 69(1), 81-92.

Önçüler, A. and Croson, R. (1998). "Rent-Seeking for a Risky Rent: A Model and Experimental Investigation", INSEAD Working Paper: 98/99/TM.

Potters, J., de Vries, C. and van Winden, F. (1998). "An Experimental Examination of Rational Rent-seeking", *European Journal of Political Economy*, 14, 783-800.

Schmidt, D., Shupp, R., Walker, J. and Ostrom E. (1993). "Playing Safe in Coordination Games: The Roles of Risk Dominance, Payoff Dominance, and History of Play", *Games and Economic Behavior,* 42: 281–299.

Shogren, J. and Baik, K. (1991). "Reexamining Efficient Rent-Seeking in Laboratory Markets", *Public Choice*, 69(1): 69-79.

Shupp, R. (2000). "Single versus Multiple Winner Probabilistic Contests: An Experimental Investigation", Working Paper, Ball State University.

Tullock, G. (1967). "The Welfare Costs of Tariffs, Monopolies and Theft", *Western Economic Journal*, 5, 224-232.

Tullock, G. (1980). "Efficient Rent-Seeking", in Buchanan, J. M., Tollison, R. and Tullock, G. (eds.), *Toward a Theory of the Rent Seeking Society*, College Station, TX, Texas A&M Press, 267-292.

Table 1: Rankings of Individuals' Ticket Purchases Across Contests

Ranking of Ticket Purchases	Percent of Subjects
p≥m≥s	27%
p=m=s	21%
p≥s≥m	19%
m≥s≥p	11%
s≥m≥p	10%
m≥p≥s	9%
s≥p≥m	3%

Note: p = number of tickets purchased in the proportionate prize contest,

m = number of tickets purchased in the multiple prize contest,

s = number of tickets purchased in the single prize contest.

Figure 1: Cumulative Frequency of Ticket Purchases by Subjects

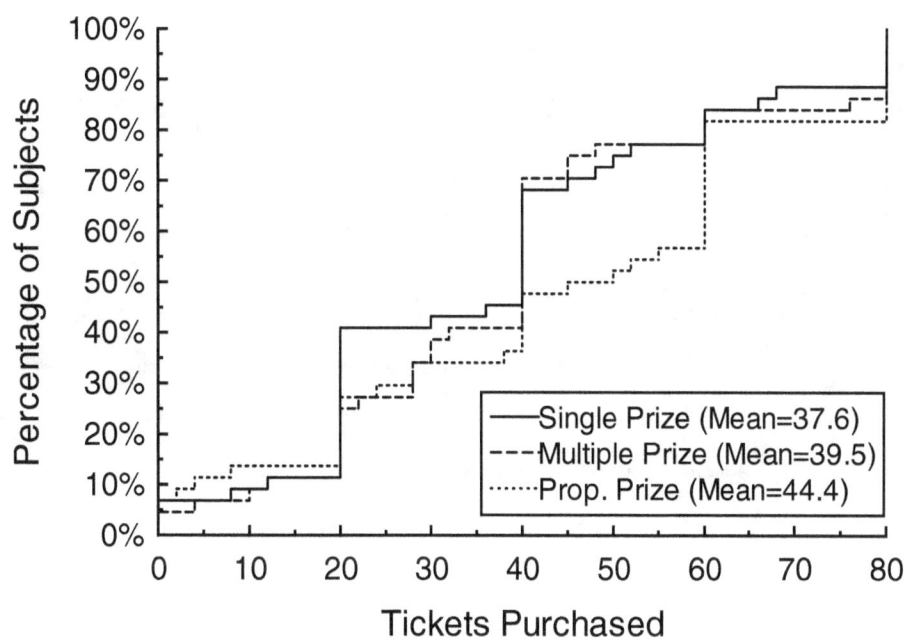

Figure 2. Correlation of Individuals' Ticket Purchases Across Contests

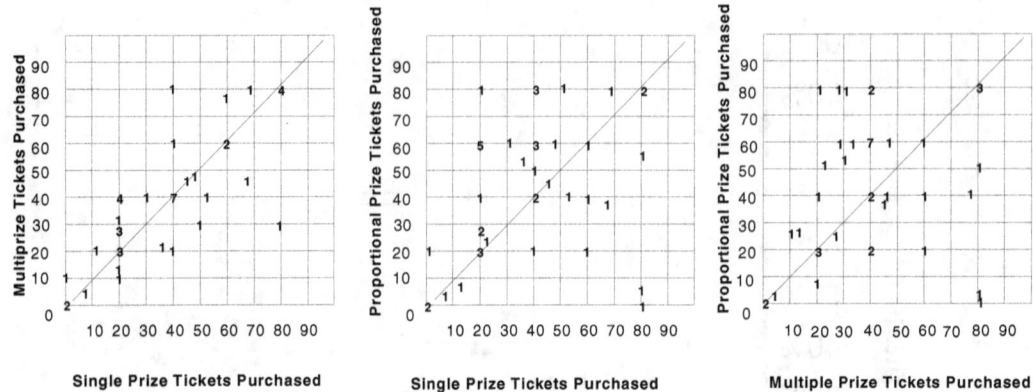

* Numbers in figures represent frequency of observations, centered on corresponding observations.

Figure 3. Comparison of Correlations Across Single and Multiple Prize Contests: Observed Actions vs. Logit Equilibrium

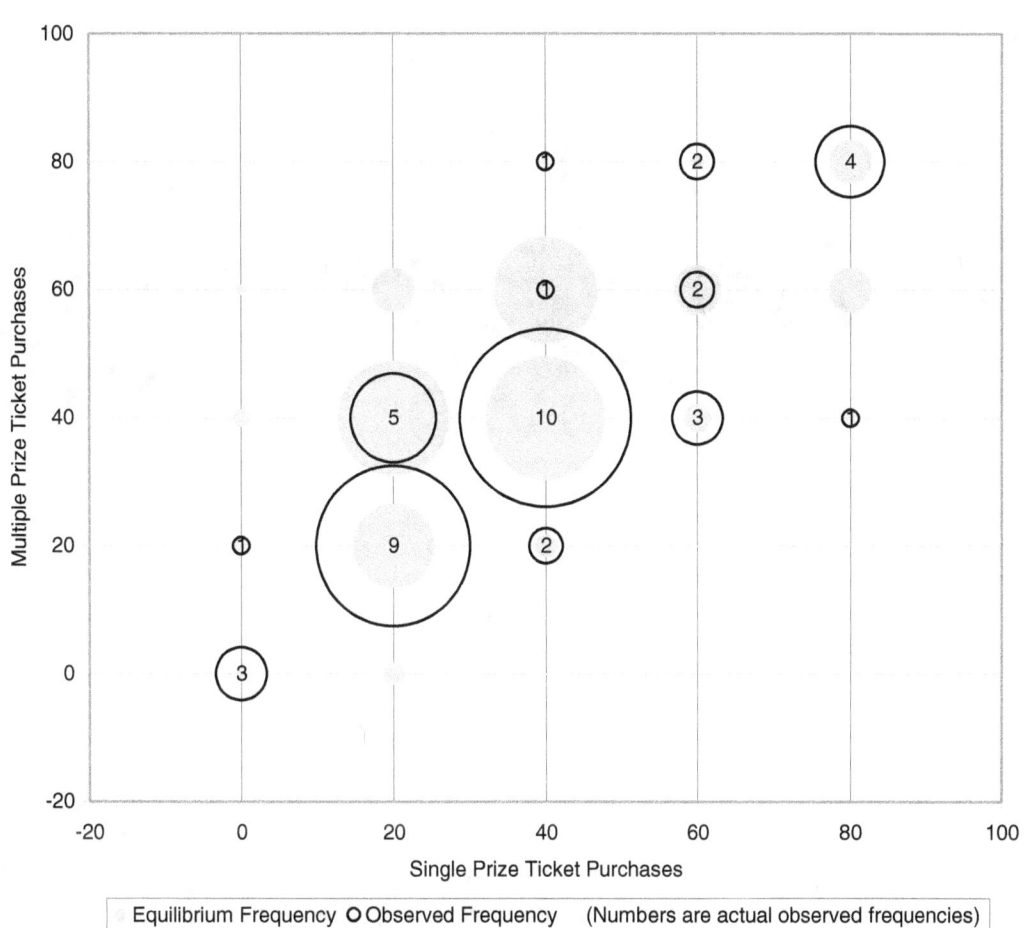

Figure 4. Comparison of Correlations Across Single and Proportionate Prize Contests: Observed Actions vs. Logit Equilibrium

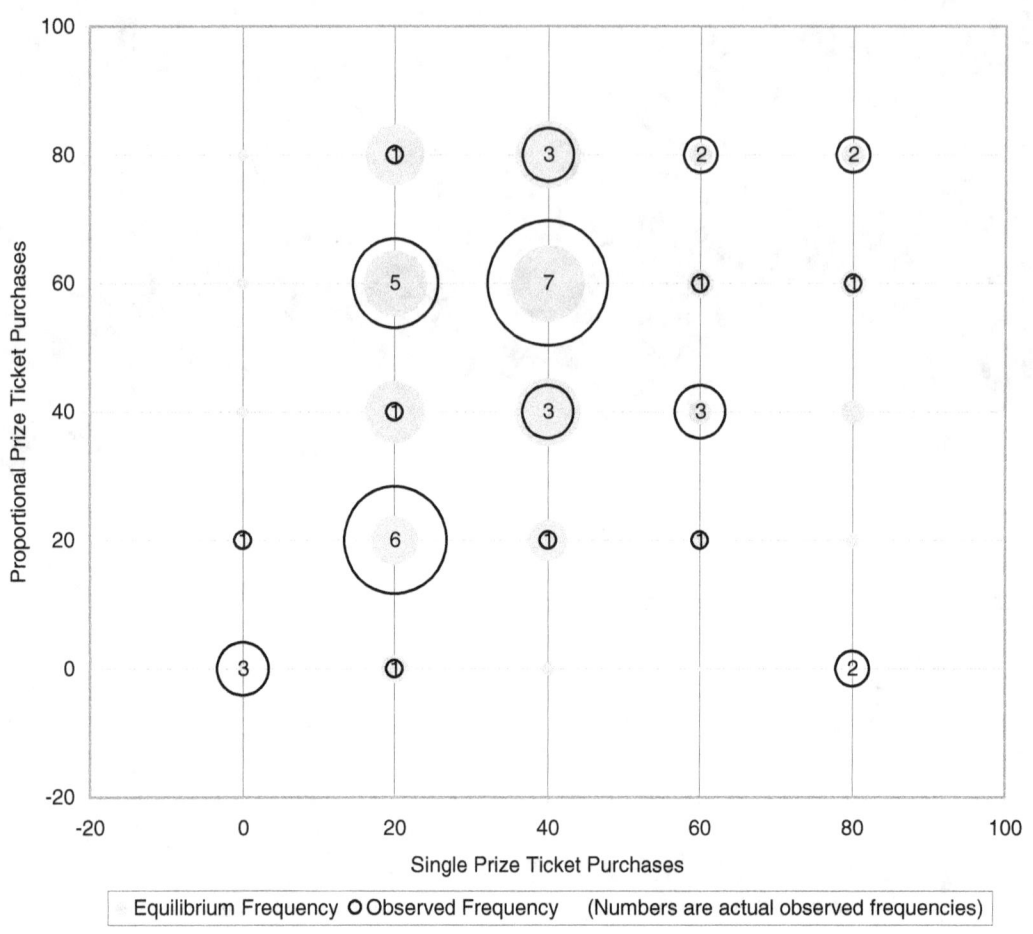

Figure 5. Comparison of Correlations Across Multiple and Proportionate Prize Contests: Observed Actions vs. Logit Equilibrium

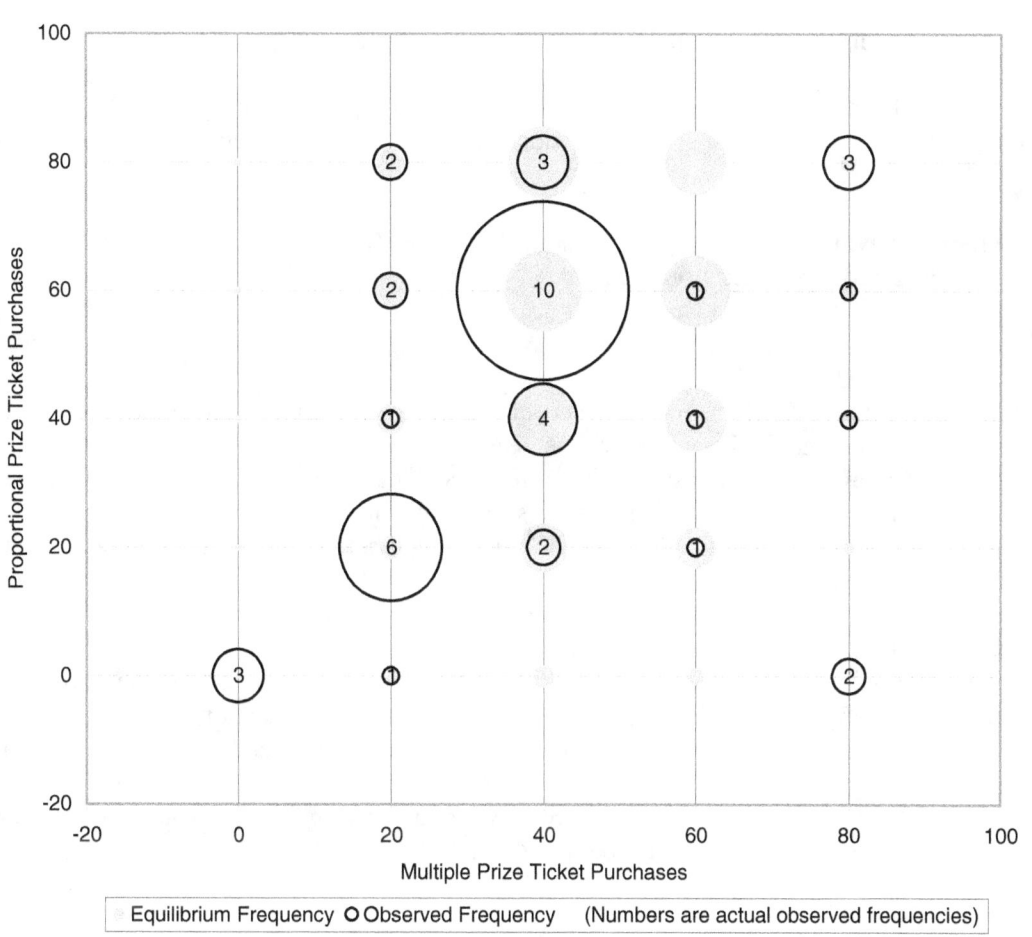

Appendix

Initial Instructions

In this experiment, you will make choices in three different decision situations.

You have been randomly assigned to a group consisting of you and three other participants. Your earnings will depend upon your decision and the decisions of the other three participants who are in your group.

All decisions will be anonymous. Further, you will never know the identities of any of the participants with whom you are matched in any decision situation.

Below we describe the six steps for the experiment.

1. You will receive instructions for the first decision situation that will explain the type of decision you will make. Then, you will be given time to make your decision in that situation.
2. We will then proceed to the instructions for the second decision situation, and you will again be given time to make your decision for the second situation.
3. Finally, we will proceed to the third decision situation, where you will be given instructions and time to make a decision.
4. Before we collect your decisions, you will be given time to go back and change any of your decisions in the three decision situations.
5. After all participants have had time to finalize their decisions, we will collect the decisions.
6. For each group, we will randomly pick one of the three decision situations. Participants in that group will receive earnings based only on the outcome of the randomly chosen decision situation.
7. At the end of the experiment, you will receive your $5 show-up fee, plus your earnings from the decision situation that was selected for your group.

Your participant number for all three decision situations is: _____

You are in group _____

Instructions for First Decision Situation

The instructions below are written to describe the way earnings will be determined if this decision situation is the one that is randomly selected for your group.

In this decision situation, one of the four participants in each group will receive a monetary prize of $72. Your chance of receiving the prize in your group depends on your decision and the decisions of the three other participants in your group.

You can affect your likelihood of receiving the prize by purchasing tickets, which you can think of as raffle tickets for the purposes of this decision situation. Your likelihood of receiving the prize also depends on the number of tickets purchased by the three other participants in your group. Prior to your decision about how many tickets you wish to purchase, you will not be able to observe the number of tickets the other participants purchase.

We will begin the decision situation by giving you $20. You can keep as much of this $20 as you like, or you can use some, or all of it, to purchase tickets.

Rules of this decision situation

You and the other participants in your group will have the opportunity to buy tickets. Each ticket will cost 25¢. Since each ticket costs 25¢, and you have $20 with which to purchase tickets, you will be able to buy up to 80 tickets.

In your group, the recipient of the $72 prize will be chosen by randomly drawing a ticket from among all of the tickets purchased by participants in your group. The probability that you receive the prize can be calculated as follows:

$$\text{Probability of you receiving the prize} = \frac{\text{(Number of tickets you buy)}}{\text{(Total number of tickets bought by your group)}}$$

If none of the players in your group buys a ticket, the prize will be awarded randomly, with each player having an equal chance of receiving the prize.

Earnings

Your earnings in this decision situation will be that part of your $20 which you do not spend on tickets, plus the $72 prize, if you receive the prize.

Demonstration

Before making a decision below, wait for the experimenter to demonstrate how this decision situation will be conducted.

Your decision

Indicate your decision in the area below:
    Number of tickets I wish to purchase: _____ (0 to 80)
    Total cost of my tickets: _____ (Number of tickets bought times $0.25)
    Portion of my $20 not spent on tickets: _____ ($20 minus total cost of your tickets)

Instructions for Second Decision Situation

The instructions below are written to describe the way earnings will be determined if this decision situation is the one that is randomly selected for your group.

In this decision situation, there will be three prizes of $24 in each group. It is possible for any individual to receive zero, one, two, or all three prizes. Your chance of receiving any one of the prizes in your group depends on your decision and the decisions of the three other participants in your group.

You can affect your likelihood of receiving a prize by purchasing tickets, which you can think of as raffle tickets for the purposes of this decision situation. Your likelihood of receiving a prize also depends on the number of tickets purchased by the three other participants in your group. Prior to your decision about how many tickets you wish to purchase, you will not be able to observe the number of tickets the other participants purchase.

We will begin the decision situation by giving you $20. You can keep as much of this $20 as you like, or you can use some, or all of it, to purchase tickets.

Rules of this decision situation

You and the other participants in your group will have the opportunity to buy tickets. Each ticket will cost 25¢. Since each ticket costs 25¢, and you have $20 with which to purchase tickets, you will be able to buy up to 80 tickets.

In your group, the recipient or recipients of the three $24 prizes will be chosen by randomly drawing three tickets from among all of the tickets purchased by participants in your group. The probability that you receive any one of the prizes can be calculated as follows:

$$\text{Probability of you receiving any one of the prizes} = \frac{\text{(Number of tickets you buy)}}{\text{(Total number of tickets bought by your group)}}$$

If none of the players in your group buys a ticket, the prizes will be awarded randomly, with each player having an equal chance of receiving each prize.

Earnings

Your earnings in this decision situation will be that part of your $20 which you do not spend on tickets, plus $24 for any of the prizes you receive.

Demonstration

Before making a decision below, wait for the experimenter to demonstrate how this decision situation will be conducted.

Your decision

Indicate your decision in the area below:

Number of tickets I wish to purchase: _____ (0 to 80)
Total cost of my tickets: _____ (Number of tickets bought times $0.25)
Portion of my $20 not spent on tickets: _____ ($20 minus total cost of your tickets)

Instructions for Third Decision Situation

The instructions below are written to describe the way earnings will be determined if this decision situation is the one that is randomly selected for your group.

In this decision situation, the four participants in your group will each receive some part of a monetary prize of $72. The share of the $72 you receive depends only on your decision and the decisions of the three other participants in your group.

You can affect your share of the prize by purchasing tickets. Your share of the prize in your group also depends on the number of tickets purchased by the three other participants in your group. Prior to your decision about how many tickets you wish to purchase, you will not be able to observe the number of tickets the other participants purchase.

We will begin the decision situation by giving you $20. You can keep as much of this $20 as you like, or you can use some or all of it to purchase tickets.

Rules of this decision situation

You and the other participants in your group will have the opportunity to buy tickets. Each ticket will cost 25¢. Since each ticket costs 25¢, and you have $20 with which to purchase tickets, you will be able to buy up to 80 tickets.

In your group, each participant's share, or proportion, of the $72 prize will be determined by the number of tickets they purchased divided by the total number of tickets purchased in their four participant group.

$$\text{Share of the prize that you will receive} = \frac{\text{(Number of tickets you buy)}}{\text{(Total number of tickets bought by your group)}}$$

If none of the players in your group buys a ticket, each participant will receive a one-fourth share of the prize.

Earnings

Your earnings in this decision situation will be that part of your $20 which you do not spend on tickets, plus the share of the $72 prize you receive.

Demonstration

Before making a decision below, wait for the experimenter to demonstrate how this decision situation will be conducted.

Your decision

Indicate your decision in the area below:
    Number of tickets I wish to purchase: _____ (0 to 80)
    Total cost of my tickets: _____ (Number of tickets bought times $0.25)
    Portion of my $20 not spent on tickets: _____ ($20 minus total cost of your tickets)

www.ingramcontent.com/pod-product-compliance
Lightning Source LLC
Chambersburg PA
CBHW081818170526
45167CB00008B/3452